Thamyres Sabrina Gonçalves

Manual for Recovering Degraded Areas in Riparian Forests

Thamyres Sabrina Gonçalves

Manual for Recovering Degraded Areas in Riparian Forests

Examples and Case Studies

ScienciaScripts

INDEX

INTRODUCTION

Environmental quality and its deterioration have been gaining public attention for centuries, according to bibliographic records dating back to slave-owning Brazil. However, the first legal milestone in this regard was only in 1934, with the publication of the first Brazilian forestry code (Federal Decree No.⁰ 23.793), which provided for the need to conserve natural environments.

Three decades later, Federal Law No. 4.771/1965 was ratified, establishing natural areas to be protected in rural areas throughout the country, such as 5-meter strips of permanent preservation on the banks of watercourses. Despite various legal modifications and additions, the protection of natural areas defined in the forestry code remains timid to this day.

It is common knowledge that, in rural areas, permanent preservation areas are often not respected in accordance with the law, which also applies to legal reserves. In 2012, the 1965 code was repealed by Federal Law 12.651, which provides mechanisms for the application and recognition of APPs, restricted use areas and legal reserves.

Among the recently instituted innovations is the Rural Environmental Registry (CAR), which brings together, in an electronic system linked to maps, the recognition and formalization of protected natural areas and other existing areas on rural properties throughout Brazil.

The idea is for the CAR to be a dynamic database, capable of integrating environmental information on rural properties and possessions, guiding actions and projects to improve production processes, socio-economic planning, as well as environmental control and land-use planning. In short, a series of lines of action, projects and programs will be able to benefit from the information that will be stored in the CAR.

However, due to the inherent complexity of the legislation and the CAR proposal, coupled with the considerable volume of properties to be covered, it is essential to plan for the adoption of standardized understandings and procedures. Otherwise, there is a great risk that the protected natural areas identified in the CAR will not actually materialize, and that the registry as a whole will not be conducive to the implementation of environmental policies, including the recovery of degraded areas.

Permanent Protection Areas in Riparian Forests

Areas of permanent preservation (APPs) are places located in rural or urban areas, whether or not they are covered by native vegetation, with the environmental function of preserving water resources, the landscape, geological stability and biodiversity, to facilitate the gene flow of fauna and flora, protect the soil and ensure social well-being. The concept of APP applies to any place that has the differentiated attributes provided for in the legislation, which mainly refer to watercourses, springs, footpaths, reservoirs, islands, hilltops, slopes with a gradient of more than 45° and the edge of a plateau.

For watercourses, the APP requirement only applies when the flow regime is perennial or intermittent. This means that there is no APP requirement for dry drainages, where there is ephemeral runoff, which only occurs during or immediately after rainfall.

The minimum strip of APP to be maintained varies mainly according to the width of the watercourse, but also when there are adjacent areas of consolidated use (Tab.1). As for the width of the watercourse, it is defined by the natural swell formed by the water, even if the level is not completely filled. When sizing the drainage channel, it is advisable to note that where water flows, there is usually no vegetation, or only sparse plants, or plants of different species from the immediate surroundings; in short, the environment itself usually expresses itself differently where there is regular drainage.

TABLE 1. Minimum strip of permanent preservation area (APP) depending on the width of the watercourse and consolidated use

WATERCOURSE WIDTH	Natural vegetation	MINIMUM APP RANGE			
		Consonant Use			
		Up to 1 module	Up to 2 modules	Up to 4 modules	More than 4 modules
Up to 10 m	30 m	5 m	8 m	15 m	Variable
10 to 50m	50 m	5 m	8 m	15 m	Variable
50 to 200 m	100m	5 m	8 m	15 m	Variable
200 to 600 m	200m	5 m	8 m	15 m	Variable
Over 600 m	500m	5 m	8 m	15 m	Variable

Source: Federal Law No. 12.651/2012; State Law No. 20.922/2013.

For properties with more than 4 fiscal modules and up to 10 modules, there is a 20 m strip of APP over the area of consolidated use. For properties with more than 10 fiscal modules, the APP strip must be half the width of the watercourse, with a minimum of 30 m and a maximum of 100 m, regardless of the location of the area of consolidated use.

It is worth noting that when there is a variation in the area protected depending on the size of the property and the presence of areas of consolidated use, the size of the property up to July 22, 2008 should be taken as a reference, in accordance with Federal Decree No.[0] 7.830/2012 and the Minas Gerais Forest Code (State Law No. 20.922/2013). It is also worth noting that these dimensions always refer to the concept of rural property; according to Federal Law 8.629/1993, this is "rural property of continuous area, regardless of its location, which is or may be intended for agricultural, livestock, plant extraction, forestry or agro-industrial exploitation". It is thus a concept associated with the notion of a unit of economic exploitation through agricultural activities, made up of one or more rural properties or possessions. And in the context of the CAR and environmental recovery actions, the rural property must be linked to the same holder or group of holders, according to MMA Normative Instruction 2/2014.

Springs and perennial waterholes are other situations that constitute APP, with a minimum radius of 50 m around them. If there are consolidated rural areas in this strip, the minimum APP radius becomes 15 m. As for springs and intermittent waterholes, these situations are not considered APP, but at the same time, the conversion of new areas for alternative land use in their surroundings is not allowed, also within a radius of 50 m and with the exception of cases in which intervention in APP is allowed. And, considering the environmental fragility of these sites, the APP use regime applies when there is natural cover and based on the precautionary principle. With the same caution, if environmental damage is identified in this type of situation, the adoption of mitigating measures should be recommended, with the support of article 16 of the Minas Gerais Forestry Code.

In any APP, intervention can only be authorized by the competent environmental agency in cases of public utility, social interest and/or for occasional activities or those with a low environmental impact. To this end, these interventions must be duly characterized and justified in a specific administrative procedure, respecting the particularities provided for in the legislation, which apply mainly to occasional activities or those with a low environmental impact:

a) Opening small access roads for people and animals, their bridges and pontoons;

b) Implementation of installations necessary for the collection and conduction of water and treated effluents, provided that the regularization of the use of water resources or intervention in water resources is proven;

c) Implementation of trails for the development of ecotourism;

d) Construction of a boat launching ramp and small mooring;

e) Construction of housing for family farmers, remnants of quilombola communities and other extractive and traditional populations in rural areas;

f) Construction and maintenance of fences, firebreaks and rainwater storage basins;

g) Scientific research into environmental resources, subject to other requirements laid down in the applicable legislation;

h) Collection of non-timber products, such as seeds, nuts, leaf litter and fruit, as long as they do not involve endangered species or species immune from logging, for the purposes of subsistence, seedling production and recovery of degraded areas, respecting the specific legislation on access to genetic resources, as well as the international treaties for the protection of biodiversity to which Brazil is a signatory;

i) Planting native species that produce fruit, seeds, nuts and other plant products, as long as it does not suppress existing vegetation or harm the environmental function of the area;

j) Agroforestry and sustainable, community and family management, including the extraction of non-timber forest products, provided that they do not de-characterize the existing native vegetation cover or harm the environmental function of the area;

k) Opening a road for reconnaissance purposes and technical and scientific surveys;

l) Undertaking de-silting and maintenance activities on dams, provided that there is proof of regularization of the use of water resources or intervention in water resources;

m) Another similar action or activity recognized as occasional and of low environmental impact in an act of the National Environmental Council or the State Environmental Policy Council (COPAM).

On smallholdings or family-run rural possessions, carrying out some occasional activities or those with a low environmental impact on APP depends only on a simple declaration to the competent environmental agency, in the case of properties already registered in the Rural Environmental Registry and practices that really cause a reduced or insignificant environmental impact. In practice, this means that, for certain activities, this simplified procedure would not be enough, such as when the interventions involve installations for collecting and conducting water and treated effluents, scientific research into environmental resources, among other situations that must follow their own protocols.

For smallholdings or rural family possessions, as well as peoples and communities considered to be traditional, the planting of short-cycle temporary and seasonal crops is permitted on the strip of land in the APP that is exposed during the period when rivers or lakes are flowing. However, this possibility must not lead to the suppression of areas of native vegetation, nor harm the quality of water, soil or wildlife.

With regard to consolidated rural areas, they are equivalent to that part of the rural property or possession where anthropogenic occupation took place before July 22, 2008, where agroforestry activities are currently carried out, where the associated buildings and internal roads of a property are accommodated. Agrosilvopastoral activities include land use practices developed, in conjunction or in isolation, for agriculture, aquaculture, livestock farming, forestry and other forms of exploitation and management of fauna and flora, mainly for economic purposes and including fallow land.

It is worth noting that when these activities are carried out in areas where the natural vegetation was converted after July 22, 2008, such alternative land use sites are not considered consolidated use in the eyes of the law. This understanding is independent of the way in which the original vegetation was suppressed, whether with prior authorization or irregularly.

Furthermore, altered areas without agroforestry use, such as dirty, abandoned pastures with low nutritional value, as well as degraded areas where there are accentuated erosion processes, widespread exposure of the soil, among other situations that ignore the use of the land, do not qualify as consolidated use.

Clearly good agronomic practices, regardless of the time of suppression. State Law[0] 20.922/2013 even imposes restrictions on the suppression of areas with native vegetation when this type of situation arises:

Art. 68: The conversion of new areas for alternative land use is not permitted on rural properties that have abandoned or unused areas.

Sole Paragraph. For the purposes of this Law:

I - the area not effectively used is that defined under the terms of a joint act by the State Secretariat for the Environment and Sustainable Development (SEMAD) and the State Secretariat for Agriculture, Livestock and Food Supply (SEAPA), with the exception of fallow areas and areas unsuitable for agroforestry activities;

II - abandoned area means the production space converted to alternative land use without any productive exploitation for at least thirty-six months and not formally characterized as fallow land.

When risks of worsening erosion processes or flooding are identified in a water APP, whether or not associated with adjacent consolidated areas, the government must determine the adoption of mitigating measures to guarantee the stability of the land and the quality of the water, subject to COPAM deliberation. Another aspect which, depending on the situation, can be assessed in a complementary manner is the agronomic suitability for the productive purpose employed. After all, depending on local conditions, activities may be being carried out that are incompatible with the natural suitability of the land and the level of technology available, which is a way of contradicting good agronomic practice and, consequently, the forestry code.

One particularity in relation to this category of area applies to properties in the integral protection conservation units in Alto Jequitinhonha, in the area of operation of the IEF regional office. The legislation defines that, for integral protection conservation units created before May 25, 2012, existing rural properties are not allowed to have consolidated rural areas, unless there is an approved management plan guiding otherwise, which does not apply to the regional reality.

When consolidated rural areas overlap the original APP strip, the legislation tolerates the continuation of agroforestry practices, ecotourism and rural tourism, as long as a minimum strip of vegetation is maintained and/or good soil and water conservation techniques are employed. On the other hand, when there is a consolidated rural area in the original strip of APP, especially where practices are carried out that are not interpreted as occasional or of low impact, there can be no suppression of new natural areas of the property for economic use, in accordance with article 20 of the Minas Gerais forestry code. In these terms, it is expected that there will be cases in which it will be more advantageous for the property owner to give up the rural area consolidated in the original APP strip in order to "recover" the right to expand agricultural activities on natural areas without differentiated limitations on their use, through an appropriate administrative process.

Environmental recovery in water APPs

In various parts of the Diamantina/MG region and neighboring municipalities, many watercourses no longer have riparian forest, erosion and silting processes are common, in short, a series of environmental values are being depreciated. And environmental degradation is inversely proportional to the productive capacity of the land, so the tendency is for these irregular situations to make life increasingly complicated for families if the panorama doesn't change. However, such a necessary change is shaped by different factors.

One limitation is that the degraded areas are often on rural properties whose owners or occupiers don't have the financial resources to invest in environmental recovery, which also applies to the IEF. In addition, many local residents still don't understand the negative effects of their actions on the value of their rural properties, the hydrological regime and the productivity of their land. Finally, in order to recover the degraded areas in these places, there is a need for parallel educational work and the ideal would be to take advantage of the same opportunity to promote better agricultural practices and other productive models (SAFs, for example), capable of improving local quality of life and preventing rural exodus.

In addition to the forestry code, Federal Decree[0] 7.830/2012 explores the expected cases of rural properties that will have degraded areas, pending regularization under the forestry code. In the context of the Rural Environmental Registry, Environmental Regularization Programs (PRAs) are planned, which will combine a set of actions or initiatives to be developed in order to comply with the forestry code. The PRAs will work through different instruments: the CAR itself, a term of commitment, a project to restore degraded and altered areas, and Environmental Reserve Quotas, where appropriate.

For landowners or squatters who adhere to the PRA, administrative sanctions for the facts that led to the signing of the commitment agreement may be suspended, in accordance with articles 59 and 60 of Federal Law[0] 12.651/2012. Therefore, the "new" forestry code does not provide universal and unconditional amnesty for infractions prior to July 22, 2008, as it depends on joining the PRA, after registering the property in the CAR, and complying with a term of commitment. As soon as this agreement is signed, sanctions arising from irregular suppression of vegetation in APP will be suspended, provided that the obligations agreed for environmental regularization are met, within the established deadlines and conditions. In this way, the penalties and fines will be converted into services to preserve, improve and recover the quality of the environment, as defined in the PRA.

The Environmental Regularization Programs are yet to be better defined in terms of their implementation and regulations. The fact is that there is still progress to be made in the procedures and techniques for applying environmental legislation within the framework of the CAR; one example is the various incentives established by law to support the recovery of degraded areas, nature conservation on private land, the improvement of environmental quality and agroforestry practices. This applies, for example, to the Mais Ambiente Brasil Program, which is coordinated by the Ministry of the Environment (MMA), according to a specific chapter of Federal Decree 8.235/2014. Other possibilities in the same direction can be found in Federal Law No. 12.854/2013, which is dedicated to fostering and encouraging actions for environmental recovery and the implementation of agroforestry systems (SAFs).

CHAPTER 1: SOIL TREATMENT AND ELIMINATION OF DEGRADATION FACTORS

Among the main aspects of recovering ecosystem functions in degraded areas is the importance of applying soil recovery techniques. For a long time, the recovery of areas was based solely on planting species, a perspective which, according to Calegari (et al, 2013) has a lot to do with the concept of community ecology based on the classical paradigm of ecosystems being closed communities, thus disregarding the dynamic process of species entering and leaving the community through the various processes to which an ecological system is subjected over different scales of time and space.

Currently, the implementation of restoration projects has gained the contribution of an important area that has gained increasing space and importance in the recovery of degraded areas, soil bioengineering, which aims to combine conventional techniques for the recovery and restoration of degraded ecosystems with methods for recovering the stability and physical quality of the soil (SCHMEIER, 2013).

It is important to emphasize that the elimination of soil degradation factors is often even more important than the recovery of the biological diversity of species, since a soil without structure will not provide conditions conducive to the development and establishment of plants or the restoration of the soil's ecological functions in the maintenance of organic matter and biological activity in the edaphic ecosystem.

Recovering soils by eliminating or containing the degradation factor focuses on rehabilitating the soil's physical structure, which is generally compromised in disturbed or degraded environments. There are various soil bioengineering techniques and the choice of the correct measure will always depend on an analysis of the levels of compaction, erosion, aeration, humidity and the granulometric structure of the soil in each area.

The environmental analysis of the areas visited in the Pico do Itambé State Park identified two areas in need of bioengineering techniques, using different soil treatment options for each:

Site 01: river bank

In this area, the main soil degradation factors identified were the intensification of the erosion process due to the removal of the riparian vegetation cover and soil compaction due to cattle trampling as a result of grazing activities without proper management. In this context, the alternatives suggested for the purpose of treating the soil and eliminating the degradation factors is to build a slope to contain the sand that is being carried into the river, which is eroding it extensively. This can be done using materials available on the property, such as wood from the trunks of trees that have already been felled or dead trees, which can be used in the stretches where erosion is occurring to a greater extent, In some stretches of the river in this same location, it is possible to contain the erosion process by planting species with a root system that favors the plant's fixation in the sandy substrate and can contain the sand being carried into the river, such as the grasses already present on the site and also some legumes (fabaceous) identified regenerating naturally on the site (*Platypodium elegans* and *Peltophorum dubium)*, Franco (et al, 1992) discuss the contributions of this family to the recovery of degraded soils. Another essential measure for recovering the soil's physical structure, which in turn will improve its chemical conditions, is to remove the cattle from the site and isolate the area from animal access.

Figure 3.1. Soil degraded by compaction due to excessive cattle trampling (a) and soil degraded by erosion (b).

Site 02: flooded lowlands

At this site, the alteration to the soil is greater in terms of its chemical structure than in terms of its physical stability, so that previous mining activities have affected its nutritional quality, in turn affecting the natural regeneration of the vegetation, which is characterized by a very small number of regenerating plant species and individuals at the site. In order to recover the physical and chemical quality of the soil at this site, it is suggested that diversified techniques be applied to induce the acceleration of the natural regeneration process, such as planting pioneering Fabaceous species that help to fix carbon and nitrogen in the soil, as well as enriching the edaphic biota, remembering the need to observe the species' tolerance to the site's periodic waterlogging conditions (see Wadt el al, 2003). The implantation of a soil seed bank from other mining areas with the same lithological structure and vegetation cover is also an interesting alternative, which has been widely used in areas degraded by mining, currently also known as topsoil (MIRANDA-NETO et al, 2010).

CHAPTER 2. DEFINING RECOVERY TECHNIQUES ACCORDING TO SITE CONDITIONS

Interventions to restore degraded areas can be carried out with different objectives, always starting with an assessment of the area's conditions, so that difficulties can be identified and strategies devised. The degradation factors and the self-regenerative potential of the areas are taken into account, based on the history of use and proximity to the source of propagules (RODRIGUES and GANDOLFI, 2001; RODRIGUES, 2002).

Another aspect to be observed is the occurrence of natural vegetation, where there may be plant and seed banks that can serve as a source of propagules for the area to be reclaimed. Kajeyama and Gandara (2001) note that the occurrence of such situations will determine the degree of intervention and the system to be adopted.

The choice of techniques to be used in a restoration project depends on the study and observation of various factors and varies according to the different situations found after the disturbance. Pereira (2009) recommends that the selection of the most suitable method(s) should be based on a detailed diagnosis of the area to be restored (figure 4.1).

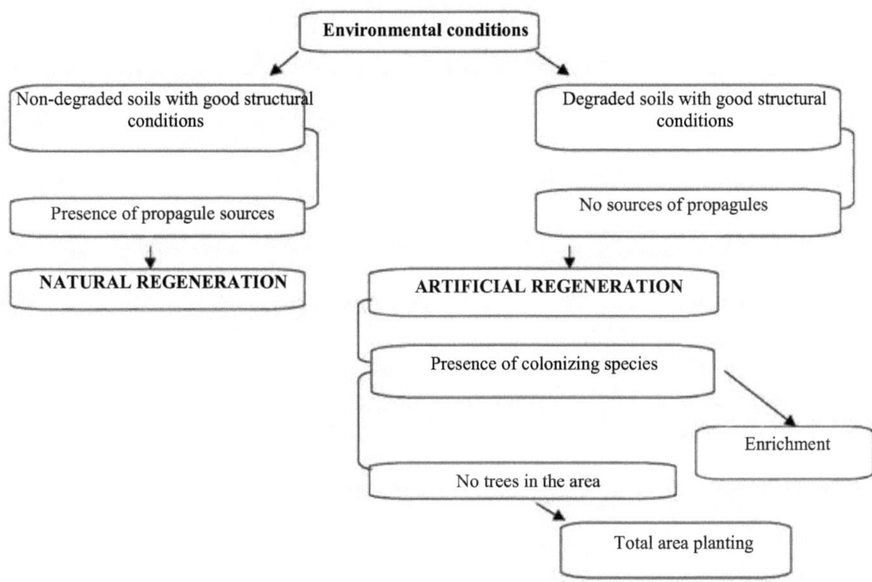

Figure 2.1: Flowchart for defining the recomposition method to be adopted based on the site's conditions. Adapted from Pereira (2009).

Considering that riparian environments are dynamic and constantly changing, the selection of the most suitable method(s) must also take into account certain factors such as those relating to the area (or site), the permanent preservation strip, the physical characteristics of the site and the species to be chosen. For example: Physical characteristics of the watershed: soil (land use, fertility, erodibility, depth and humidity); hydrology (extent of floodable areas and average duration of flooding periods) and topography (slope).

Generally speaking, the recommendations for restoring these areas will basically depend on the degree of disturbance and can be as follows:

1.   In situations where there is a low probability of accelerated erosion, with the presence of colonizing plants and close to remnant forests (source of propagules), only superficial soil protection can be carried out, providing conditions for natural plant succession (RODRIGUES, 2002). Priority should be given to species with both shallow (grasses and legumes) and deep (medium-sized trees and shrubs) root systems;

2.   In situations where there is a potential for accelerated erosion processes and/or where environmental recovery and control of erosion processes already in place are necessary, or when it is necessary to set up massive populations (on the edges to protect partially affected forests, creating clumps linking remaining forests, etc.), fast-growing species should be used.), fast-growing species should be used, observing the dominance of plant associations that occur in the region, maximizing the use of native species, even if they are pioneers, with the aim of recolonizing the flora and fauna (RODRIGUES and GANDOLFI, 2001);

3.   In situations involving the restoration of riparian forests on the banks of rivers and varzeas, the solutions to the problems of both containing erosion processes and recovering the continuity of riparian formations, with their specific floristic diversity, should be highlighted;

4.   When restoring areas around springs, it is necessary to fence off the area within a minimum radius of 50 m from the waterhole to prevent the presence of domestic animals such as horses and cattle that use the area to obtain water. As well as feeding on some of the seedlings of native species in regeneration, these animals can trample on the seedlings, preventing them from establishing; contaminate the spring water; compact the soil, making it difficult for water to infiltrate (PEREIRA, 2009);

In short, the decision on the most appropriate way to restore the environment will depend on an analysis of the local situation and knowledge of the ecosystem. Restoration techniques range from those that require no direct intervention to those that are highly interventionist. Non-interventionist techniques are basically related to eliminating the source of degradation and depend on landscape features that can favor the natural regeneration of the degraded area, such as the proximity of remaining forests. Intervention requires more direct actions, such as direct seeding and planting seedlings of forest species, as well as eliminating the barrier to regeneration (usually another plant considered invasive, such as some grasses). However, the level of intervention of the techniques adopted varies widely, as suggested in Table 4.1 (RODRIGUES & GANDOLFI, 2000).

TABLE 2.1. Proposed actions for the restoration of degraded areas, with different levels of intervention.

Restoration actions	Principles and constraints
Area isolation	prevent further degradation; local resilience must be preserved
Removal of degradation factors	correctly identify the agent of degradation; strong regeneration potential
Selective elimination of competing species	when there are unbalanced populations of species that inhibit natural regeneration
Enriching species with seedlings or seeds	planting or sowing where there is low plant diversity and little dispersion
Implementing a consortium of species using seedlings or seeds	planting or sowing in places where there is no forest or remaining seed bank
Induction and conduction of autochthonous propagules	induction and conduction of existing propagules (rain or seed bank)
Transplanting seeds or seedlings	transfer of seed bank (litter) or seedlings to degraded site
Use of interactions between plants and animals	attraction of dispersing animal species, with the aim of facilitating the succession or planting of mycorrhizal species, e.g.
Planting economic species	use of species with economic potential (timber, honey, fruit) as an alternative source of income

Source: adapted from Rodrigues & Gandolfi, 2000.

In the surroundings of the Pico Do Itambé State Park, MG, for example, mining, sand and gravel extraction and the formation of pastures for cattle breeding have triggered degradation processes, especially erosion processes; with varying degrees of severity in a large part of the length of the Capivari River, in nearby springs and streams (figure 4.2).

Figure 2.2. Erosion processes, with different degrees of severity, found around the Pico do Itambé State Park, in the municipality of Serro, MG. (Source: Lourenço, 2014)

In general, the planting of ground cover using poaceous and herbaceous fabaceous species; the planting of tree species seedlings; the conduction of natural regeneration; the use of direct seeding and measures to activate ecological succession are methods that can be used effectively to recover the vegetation of riparian permanent preservation areas and are presented below.

2.1. Ground cover planting: use of poaceae and herbaceous fabaceae species

The stabilizing function that vegetation exerts on the soil, preventing or minimizing erosion processes, is achieved through a combination of mechanical and hydraulic factors (GYLSEN & POESEN, 2003). Mechanical reinforcement is mainly carried out by roots, by anchoring surface layers to more stable and deeper layers (MATTIA et al., 2005). Roots affect soil properties such as infiltration rate, aggregate stability, moisture, organic matter and shear strength (GYLSEN & POESEN, 2003). Roots also hold surface soil particles during runoff and increase soil roughness, providing high infiltration capacity and reducing surface flow velocity (De BAETS et al, 2007,

GYLSEN & POESEN, 2006).

Implementing a recovery method does not necessarily involve planting seedlings (PEREIRA, 2009). For the recovery of degraded soils, green manure is recommended, i.e. the use of one or more plant species that fulfill the function of protecting and nourishing the soil, generating better growth conditions for other plants, accelerating natural regeneration and reducing soil loss. The function of these plants is to produce organic matter to cover the soil, protecting it from the sun and rain, and to provide energy and nutrients for soil organisms, improving the soil's physical, chemical and biological conditions.

Green manure species are generally annual herbs from two families: grasses and legumes, i.e. poaceae and fabaceae, respectively. Grasses are recommended for maintaining and improving soil properties and mitigating erosion processes, due to their rapid growth and efficient soil cover (De Baets et al. 2006). De Baets (2006) proved that the roots of grass species increase the shear strength of the soil and resistance to water erosion. In addition, according to Stokes et al. (2009), the presence of grass roots is also effective in mitigating and preventing surface erosion due to their association with fungi and bacteria that increase the cohesion of soil particles. Legumes are also recommended (ANGIE & HAU 2009). They are able to fix nitrogen in the soil by associating with bacteria that coexist in their roots. Grasses, on the other hand, have a high carbon content. Therefore, a consortium of these two families is recommended in order to maintain a good carbon/nitrogen ratio.

However, despite the wide diversity and distribution of native grasses and legumes in Brazil, most of the work aimed at restoring and revegetating degraded areas uses exotic species (MARTINS, 2006). The study of native grasses is fundamental to understanding ecological processes in natural environments, as they are able to successfully replace exotic and invasive species, which compromise the ecological balance of ecosystems (PIVELLO, 2005).

Currently, the lack of basic knowledge about the germination, establishment and development rates of native species is undoubtedly the biggest impediment to the use of native species in revegetation processes. Such use could make a major contribution to maintaining local biodiversity, promoting the protection or even expansion of natural sources of genetic diversity, as well as providing advantages such as lower production and transportation costs for seedlings (MOREIRA, 2002). It is suggested that research be carried out into techniques for sowing native species, as an alternative to the use of brachiaria and fat grass, the aggressiveness of which can damage natural environments.

In a field study on a private property located in the municipality of Serro, in the area surrounding the Pico do Itambé State Park, we visited an area with gently undulating relief, on a plateau, originally covered by cerrado and with rocky fields on relatively nearby rocky outcrops. This is the bank of a stream, where the soil is sandy and used for grazing, which has led to processes of bank erosion and silting up, with possible modification of the watercourse as a result of these processes (figure 4.3). It is worth mentioning that this stream drains into one of the most important natural attractions around the park, in the Capivari region, the Tempo Perdido waterfall. Part of the surrounding area is occupied by cerrado (savannah) in the midst of brachiaria. At specific points on the banks of the stream there is good natural regeneration of candeias (figure 4.3).

Figure 2.3. Partial view of a site to be reclaimed, with a stream bank washed away and a brachiaria APP used as pasture, in the surroundings of the Pico do Itambé State Park, Serro municipality (Source: Lourenço, 2014).

2.2. Planting tree seedlings to restore riparian areas

The implantation of tree species is a procedure that allows the initial stages of natural succession to be skipped, where herbaceous and grass species first appear, enriching the soil with organic matter and altering its characteristics, thus allowing the appearance of shrubs and trees. In heterogeneous planting with regional native species, shrubs and trees can be planted simultaneously, making it possible to accommodate both pioneer and non-pioneer species (RESENDE, 2006).

Seedlings can be planted on the whole area or on part of the area, using the densification or enrichment technique. An important factor to consider when choosing this method is the cost of implementation and the availability of seedlings. However, it has a huge advantage in terms of the survival rate of the seedlings in the field and the possibility of choosing the species (or ecological groups) to be planted in the area, which can guarantee the success of the restoration.

The species selected for each site should be those that occur naturally in climate, soil and humidity conditions similar to those of the area to be restored. The differential adaptability of the species to each environmental condition identified in the riparian strip should be taken into account. The selection of species capable of inducing a new resilience can be based on the choice of: pioneers and early secondaries (species that grow in full light); late secondaries and climax (slow-growing species, developing better in the shade). It is also advisable to choose species that specialize in nourishing the soil through symbiotic processes with nitrogen-fixing bacteria and mycorrhizal fungi. It is also necessary to try to involve different pollination and seed dispersal syndromes in order to ensure that animals can be present in the area throughout the year (REIS, *et al.*, 1999). The possibilities and pretensions of restoration projects for degraded riparian areas where the regional matrix is still forested are very different from those proposed for places where the matrix is no longer forested. In the first case, restoration basically depends on creating the necessary conditions for the arrival and establishment of propagules from the surrounding forested areas. In situations where the regional matrix also consists of heavily altered or degraded areas, the riparian restoration project itself must include strategies that guarantee the establishment and perpetuation of the natural characteristics of the restored vegetation, such as high biodiversity and complexity of interactions, not to mention the possible contributions of propagules and genetic variability from neighboring areas (RODRIGUES & GANDOLFI, 2000).

Generally speaking, it is recommended that native pioneer species be kept in the area, and that certain native species be introduced to enrich local biodiversity and control invasives, according to the ecological vocation of each species (species found in wet, sloping areas, forest margins, etc.).

2.2.1 Selection of species used

Riparian forests provide the ecosystem with a range of benefits in terms of its many hydrological functions. Moisture is a determining factor in the distribution of species and generates strong seeding pressure, which requires the presence of species that are well adapted to these environments. The constant flooding in the areas of infiuence of the riparian forests is one of the main factors in the separation of species that have developed adaptive strategies for these ecosystems (PINTO et al., 2005).

It is important to know that riparian forests comprise different environments, ranging from mesic sites, with no influence from floods, to areas of deposition, where the plants are partially or totally submerged. Therefore, it must be understood that the appropriate selection of species is the key to the success of the planting (DAVIDE et al., no date).

In order to select the species used, a fioristic inventory must first be carried out, the function of which is to infer the behavior of the species in the community. If it is not possible to carry out a detailed inventory, we suggest carrying out quick surveys of the nearby remnant vegetation in order to obtain information that indicates the most suitable species for the different microsites and for planting in the various stages of degradation of the riparian forest. In addition to the most frequent species, the survey should be enriched with information on species that occur in peculiar situations such as woodlands, wetlands and riverbanks. This information will provide a better basis for choosing species and deciding at which stage of recovery they should be classified

(Embrapa, 2000).

Another important and complementary criterion for species selection is obtaining seeds of the species. Based on the species quota, the seed stock should be used for germination tests, growth of piantuias in nurseries, seedling production and direct sowing in survival experiments (reforestation methods). Potential germination and growth speed should be analyzed for the selection of species for reforestation based on the planting of seedlings in degraded environments. High germination rates reduce seedling collection efforts, while rapid seedling development minimizes nursery maintenance time and costs (figure 424), thus providing field survival in competition with invasives (figure 2.5) (SUN et. al., 1995).

Figure 2.4: Partial view of the nursery at Pico do Itambé State Park, on the São João farm, Santo Antonio do Itambé municipality. (Source: Lourenço, 2014)

The selection of native species should be evaluated according to their successional stage, growth and survival in field conditions. Classifying the Ecological Group is a strategy for differentiating species within the dynamics of forest succession and is related to the behavior of species in relation to exposure to light, which can be classified according to certain criteria: pioneers (P), early secondaries (SI), late secondaries (ST), shade-tolerant climax (CS) and light-demanding climax (CL). Today, the trend among researchers is to classify species as: pioneers, shade-tolerant climax and light-demanding climax (Napo et al).

Multivariate techniques are used to classify ecological groups. Based on association matrices, multivariate data analysis, of which there are several types, leads to different results each time. There are two alternative ways of studying vegetation:

- Cluster analysis studies using agglomeration or division algorithms based on dissimilarity measures;

- Ordination studies, using principal component analysis and correspondence analysis (SANTOS et *al.*, 2004).

Seedling planting is considered to be one of the most widely used reforestation methods, mainly because it provides a good initial density of plants. However, one of the method's drawbacks is that it reconstitutes vegetation with a uniform structure, which is very different from the heterogeneous structure of natural forests. The method can also be difficult to apply in areas with very uneven topography or steep slopes. It is suggested that the recovery process be accelerated by sowing species, which is based on increasing the potential for self-recovery by densifying and enriching the seed bank, in areas whose history indicates its absence, or in areas already occupied by pioneer species by sowing late species (LACERDA and FIGUEIREDI, 2009).

2.2.2 Composition of planting models

In order to determine species association models for use in restoring degraded areas in riparian forests, the characteristics of successional groups with different tree species and specific ecological functions must be observed (ARAGÂO, 2009).

The most common intervention in degraded areas is the use of mixed plantings of native tree species, which act as catalysts for ecological succession (PARROTTA et al., 1997). Plantations in areas of unnatural regeneration, as well as in areas of slow regeneration, have the function of accelerating the process of secondary succession, or even providing the conditions for it to occur, as they act as real perches attracting fruit and seed dispersers and thus reintroducing new species and intensifying the process

(MORAIS et. al., 2013).

The seedlings can be planted randomly or in grouped arrangements and the plants can be distributed in modules or in rows (BOBATO et al., 2008).

According to Morais et. al., in module planting, a seedling of a late secondary species is surrounded by seedlings of early secondary species, which will support the growth of the first, while a seedling of a climax species, which develops completely in shaded conditions, is surrounded by seedlings of pioneers. When planting in rows, species can be combined by alternating rows (i) with pioneer species only and (ii) with early secondary species and rows with late species interspersed between the fast-growing ones.

The conditions provided for the planting can change depending on how the plants are arranged in the field, whether in modules or in planting rows. It should be noted that row planting is more suitable for large-scale planting (tens to hundreds of hectares), when the operation is automated. Mulching is more precise and is more recommended for small-scale and experimental farms. The most important thing is to ensure the spacing and quantity of seedlings for each eco-organic group, so that each eco-organic group has the greatest chance of occupying the most suitable space possible, in the shortest time and for the longest period (KAGEYAMA, 2002).

Based on the process of secondary succession, the planting of pioneer species has the function of covering the soil and providing shade for the cyma species. In this case, quincunx planting is suggested, as it promotes the establishment of plants using pioneer species at the vertices of the square and a cymax species in the center (DAVIDE et al., 2000).

Direct sowing, which is a suitable artificial method and can be done throughout the area, is suggested for sowing in trenches or pits (BARNETT and BAKER, 1991). Seed sowing can be used to introduce initial species (pioneers and early secondaries) into areas where there is no vegetation and also for non-pioneer species (late secondaries and cymax) when working to enrich secondary forests (KAGEYAMA and GANDARA, 2004).

The use of ecological succession in the implementation of land reclamation is an attempt to promote artificial regeneration in a way that follows the conditions under which it occurs naturally in the forest. Generally, the succession model is used to separate species into ecological groups, and to group them together in such a way that the earlier species in the succession provide adequate shade for the later species (KAGEYAMA, 1984). Basically, the pioneer species provide the cyme species with more intense shade, while the early secondary species provide partial shade for the late secondary species (KAGEYAMA et al., 1990).

Agroforestry systems represent the integrated use of land and forestry, agricultural and livestock production and can be used as a recovery model for producers. The main objective of this method is to allow greater diversity and sustainability in the area and, from an ecological point of view, it allows more than one species to coexist in different niches (SANTOS and PAIVA, 2002).

Below, in figure 4.6, is an example of a site where the agroforestry model is suggested as a proposal for restoring the area.

Figure 2.6: Partial view of a stretch of degraded hillside, and the bank of a watercourse occupied predominantly by brachiaria, just below the hillside surrounding the Pico do Itambé State Park, municipality of Santo Antonio do Itambé. (Source: Lourenço, 2014)

It is a site with a strongly undulating slope, with stretches that can exceed 45^0 , constituting APP. The original vegetation was Semideciduous Seasonal Forest, but the land is currently very exposed due to illegal logging, with no adjacent forest matrices. It is a hillside, with a perennial watercourse just below, whose APP is degraded, with significant brachiaria cover.

Although planting models are a great tool for riparian forest recovery projects, they are as much a theoretical concept as an applied one, and a lot of discussion and experimentation is needed to apply them.

2.2.3 Soil preparation

The aim of soil preparation is to improve the physical conditions of the soil and/or to incorporate fertilizers and correctives in order to encourage the establishment of the stand. The equipment and techniques to be used in soil preparation will depend on the chemical, physical and topographical characteristics of the soil, as well as the availability of financial resources to carry it out (NAPO et al., no date).

a)   Planning phase: During soil preparation, you should plan how the green manure species will be planted.

b)   Installation phase: Soil preparation should begin with mowing and then rotating with the tractor, if it is possible to mechanize the area, depending on the type of terrain.

To rotate, it is first suggested that the tractor be driven in low gear, in order to decompress the soil, leaving it loose. Next, the tractor is shifted into a faster gear to level the ground and better mix the organic matter from the mowings. Once this has been done, it is also important to cover the ground to allow water to run off during rainy periods, preventing the soil from becoming waterlogged.

The next step is to plant species for green manure in order to restore soil fertility. There are several species that can be used.

Various sources of organic material added to degraded soil have been used to improve soil properties (KITAMURA et al., 2008). Alves & Suzuki (2004) analyzed that the use of cover crops combined with crop succession under direct sowing can improve the soil's physical properties, such as porosity, soil density and mechanical resistance to penetration. Aguiar et al. (2000) observed that an alternative that can speed up the recovery of degraded areas is the use of species that are native to the area, together with species that speed up the chemical and physical balance of the soil, with green manures being of great importance for rebalancing the ecosystem, as they are important for initial soil cover. The use of soil preparation methods, their variations and timing are defined after visiting the area, surveying and analyzing local conditions.

According to EMBRAPA, after clearing the area, plowing, harrowing, subsoiling and furrowing are carried out to prepare the soil. Furrowing should be done to a depth of 20 cm, with a spacing of 2 m to 3 m between the crop rows (taking into account the growth habit of the cultivars). Instead of furrowing, pits can be used as a way of breaking up compacted soil layers, which should be practiced mainly on slopes in order to prevent soil erosion. These techniques, which are generally used for annual crops of economic interest, can also be used to restore degraded areas, provided they are well assessed, planned and adapted to the reality of the situation.

Soil preparation can be divided into several stages. It depends on the type of soil and the availability of machinery and sometimes some stages are omitted because there is no equipment available or for reasons of economy, such as using harrowing instead of plowing.

2.2.4 Fighting ants

Combating and controlling ants is essential for the success of forest restoration, given the ability of these insects to damage plantations. Leafcutters, sauve (*Atta* spp.) and quenquéns (*Acromymex* spp.) are considered to be the most important pests for forestry species in Brazil and can therefore make the project unviable. It is therefore advisable to combat ants when planting the species and during planting.

This technique is carried out according to the environmental conditions of the site, the type of anthill present in the area, the level of infestation and the products and equipment available. Each product used requires the use of personal protective equipment (PPE) appropriate to its characteristics. Among the formicides found on the market, there are solid (granulated or powdered), liquid (thermo-nebulizable) and gaseous products.

The bait is currently the most widely used in the fight, as it is practical to apply, low in toxicity and cost, and also gives good results. The poison is usually packaged in plastic bags (around IOg), which are placed along the paths near the active scouts. The number of baits depends on the size of the anthill and the product selected. It is recommended that the baits be applied during dry periods.

The second technique is thermonebulization, which consists of a mixture of liquid insecticide and diesel oil, transformed into smoke and injected into the anthill by a motorized device called a thermonebulizer. The smoke that penetrates the anthill kills the ants on contact. It has the advantage of being able to be applied in both dry and rainy weather, just to control the ants. It is one of the most efficient methods of combating ants, but the most expensive.

Another alternative is dry powder, which, although low-cost and highly efficient, is labor-intensive. This method is recommended for small anthills (saùvas and quenquéns). It is applied using sprinklers, which inject the powder into the anthill, killing the ants on contact. Recommended for dry weather, as wet soil makes it difficult for the product to act.

Combatting pests: Measure the size of the anthill by going over the entire area of soil and measuring its length and width in the shape of a cross. Calculate the area of the anthill by multiplying the greatest length by the greatest width. Locate the mound of loose earth. Measure the greatest length and width. Choose six scouts at the edge of the loose soil and scrape until enough ants come out.

Apply eight doses of 6 g to each of the six scraped-out holes. This procedure will result in the correct dose being applied every time, regardless of the shape and size of the anthill. Apply a dose of 6 g of bait to all the scouts that are more than one meter away from the mound of loose soil.

For Quenquen ants: Place the ant killer next to the carriers and in a dry place in the soil. If the soil is wet, apply the ant killer to leaves or tree bark. Do not store formicide together with products that give off an odor, such as gasoline and others. Store pesticide containers away from children. Do not store pesticide containers in direct contact with the ground. Do not smoke during the application of formicide. Do not apply formicide in areas with loose soil.

2.2.5 Fertilization

Soils in degraded environments generally have a low capacity to support healthy and productive plants, mostly due to erosion which carries away the surface layers, rich in nutrients and organic matter. When left without vegetation, the degradation process can be accelerated (Brani et al. 2003).

Liming in acidic soils has been studied with the aim of correcting acidity. However, in riparian forests, the species used in riparian reforestation are well adapted to acidic soil conditions, as most of them are of secondary class. Thus, in this context, the main objective of the liming process is to increase the availability of the nutrients Ca and/or Mg for the seedlings. Lime can be applied directly to the bottom or around the seedling's planting hole, using 200 to 300 g per hole. It is recommended that the limestone be applied before planting, preferably two months before the seedlings are planted, by spreading it over the entire area.

The term "riparian forests" involves all types of tree vegetation linked to riverbanks, along with the high species diversity of the rainforest. This prevents detailed knowledge of the absorbable amount of each nutrient per species for a specific recommendation. However, it is possible to make general recommendations after analyzing the soil chemistry. Fertilization should be done to correct the most severe deficiencies in the three most important elements for plant development: phosphorus, potassium and nitrogen.

Even though the demand for nutrients varies between the different species present in the area, the season and the stage of growth, based on a soil analysis carried out on samples collected in the Pico do Itambé State Park, we recommend a generic formulation for planting seedlings in riparian forest reforestation in six areas.

Table NUMBER: Size of pit and fertilizer used in recovery planting in Itambé State Park

	Cova	T/ha of CaCO3	Starter fertilizer
Area 1	30x30x30	1.98	50g of SS per pit
Area 2	30x30x30	2.52	50g of SS per pit
Area 3	30x30x30	1.47	50g of SS per pit
Area 4	30x30x30	1.74	50g of SS per pit
Area 5	30x30x30	0.15	40g of SS per pit
Area 6	30x30x30	1.44	50g of SS per pit
Area 7	30x30x30	1.32	50g of SS per pit

In general, the sites under study have good PH conditions, good and very good levels of the main nutrients: nitrogen, phosphorus and potassium, as well as magnesium and calcium. However, there were high levels of aluminium, a toxic substance for plants, which will be corrected by liming.

Species in the early successional stages, due to their greater growth potential or rate of biomass synthesis, have a greater capacity to absorb nutrients than those in the later successional stages. It is therefore advisable to avoid the use of quick-release fertilizers such as supersimple.

The NPK formulation (20, 00, 20) can be used, applying 200 grams per plant, or equivalent, in a semi-crown, during the rainy season. Plus the application of 90 kg/ha of Potassium Chloride (or approximately 50 g/plant). In order not to encourage the growth of invasive plants, the fertilizer should be applied after cleaning such as weeding or under conditions of low weed infestation and competition.

Top dressing is recommended 90 days after planting, distributing the chemical fertilizer around the plant, avoiding a distance of 20 centimeters around the seedling. Another recommended mineral formulation is to use 350 to 500g of P2O5, 200g per plant of KCl together with micronutrients: 6kg per ha of Zn, 1-2Kg per ha of B, 1-2Kg per ha of Cu. (Resende, 2003)

2.2.6 Piano maintenance

Restoration plantations are ideally maintained until no later than the second or third year. Invasive species should be eliminated whenever necessary, until the seedlings outgrow the herbaceous vegetation. Selective clearing of the area should be carried out to eliminate invasive species, in the form of weeding in strips between the rows of plants. In addition, crowning, clearing or elimination of invasive species should be carried out around the seedling.

Manual mowing is selective and indicated for areas where mechanized mowing is not possible or in places with high regeneration potential.

Manual chemical mowing between the planting lines consists of preparing the spray mixture and applying the herbicide with a backpack spray pump around the seedling, duly protected with a bucket or PVC pipe. The need for weeding and crowning is assessed visually. It should be noted that the natural regeneration of native shrub and tree species must be preserved.

Combating ants in the maintenance phase follows the recommendations made earlier when preparing the area. It is important to carry out periodic monitoring of restored areas in order to determine the need for ant control. The mere presence of ants in the area should not be the only factor considered. A small infestation is tolerable and considered normal, as long as the damage caused by the attack is small. In areas of significant size and high species diversity, it has been found that ant control is necessary until no later than the second or third year.

In the event of mortality, replanting should be carried out. This should be done in the next rainy season, i.e. one year after the first planting in that location, in order to ensure greater survival.

2.2.7. Planting Season

Planting should be carried out at the beginning of the rainy season (November), because if it takes place outside of this time, the mortality rate of the seedlings introduced to the site increases dramatically, due to the occurrence of veranicos, which are common at this time of year.

2.4 Using direct seeding to restore riparian APPs

2.4.1. Sowing in a pit

WHEN TO USE

Suitable for very steep areas, where the slope is a limiting factor to the entry of machinery. Therefore, digging pits on contour lines allows planting and reduces losses due to drift (Sitio X).

It is widely used to enrich reforestations with late species that were not included initially, or even in places where replanting is necessary.

HOW TO USE

1. Prepare the area that will receive the seeds by weeding to contain the weeds.

2. Open the pits.

3. Add fertilizer to the bottom of the pit if necessary*.

4. Sow between 3 and 5 seeds per hole, depending on the germination characteristics of the species to be used.

SUGGESTION

If the soil is highly compacted and the availability of nutrients is negligible, it is advisable to prepare a substrate with the characteristics of interest and use it to close the hole, not returning with the same material that was removed when the hole was opened. This will allow the seedling to find slightly more favorable conditions for its initial development.

2.4.2. Row or Furrow Sowing

WHEN TO USE

This is ideal when the right machinery is available for the job, as well as when the soil has large physical limitations. Or when the aim is to recover large areas with a quick covering of the exposed soil.

Another situation in which this practice is appropriate is when you want to use SAFs (Agro-Forestry Systems) to restore the APP (SitioY). In other words, the landowner can use economically valuable crops together with the native species needed for environmental suitability, interspersing them. A row can be planted with forest species and a row with agronomic species of interest, in sequence, allowing income to be earned until the forest species are large enough to shade the area and provide the necessary conditions for ecological succession.

The use of SAFs allows income to be generated until the forest is established. It is a **very interesting** technique to propose to small producers who use the land for their own livelihood, since environmental adaptation will not immediately render the area unproductive. In this sense, it tends to encourage the community to take a greater interest in restoring riparian areas. Since a successful project is precisely one that results in techniques that suit people's different understandings, resources, goals and needs (Campos-Filho, 2013).

HOW TO USE

1.   Carry out harrowing in the planting line or on the whole area (heavy harrowing if the area has a lot of physical limitations and dead matter, otherwise light harrowing);

2.   Planting is carried out using a planter, or the furrow is opened and then planted. (The seeds can be distributed manually or using a manual or mechanical seeder. Manual sowing is more laborious, less efficient and requires more seeds. It can be done using a small can with a 4 to 5 mm diameter hole in the bottom, or a wide-mouthed glass with a similarly perforated lid. By shaking the can or glass full of seeds and with a hole in the furrow line, the seeds will fall into the furrow (Ageitec, 2015).

3.   Standardize sowing at a depth of approximately 3 cm. Seeds sown too deep have difficulty emerging and seeds sown too shallow can be desiccated or predated.

4.   It is important to establish a minimum distance between plants and between planting lines, according to the size of the forest species to be used, to allow for harmonious development.

5.   Regardless of the agricultural crop selected, it is imperative that the soil is kept covered with vegetation at all times. Beans can be used for this purpose.

SUGGESTION

Although interesting, the use of SAFs should **not** be recommended for the whole area. What is suggested is a combination of different techniques in the same area, respecting the characteristics of the specific sites present there.

2.4.3. Sowing

WHEN TO USE

Recommended for farms that already have the agricultural machinery needed to spread the seeds on the surface (seed drills or agricultural planes) and incorporate them using a harrow.

Not recommended for very steep areas, as the slope can carry the seeds to the lower parts.

HOW TO USE

1. Grading the entire area

2. Incorporate lime in total area, if necessary

3. Sowing

4. Grate lightly to incorporate the seeds

SUGGESTION

On better sites, only tree species can be sown, while on poorer sites it is advisable to mix them with seeds of fast-growing herbaceous species.

2.4.4. Muvuca de Sementes

The official name of the technique is "mechanized planting of native cerrado seeds". The idea for the name Muvuca came from the fact that various native and agricultural species are mixed together.

WHY USE IT?

It is a cheap and accessible technique. It can replace or complement the planting of seedlings when this is made difficult by the lack of nurseries in the region you want to restore or of seedlings ready for shipment at the right time.

Reforestation by muvuca seems uninteresting at first, as only around 4% of the propagules sown will grow and establish themselves.

However, even if most of the seeds sown don't come up, this type of planting is still cheaper. Sometimes the quantity is not necessarily what indicates that one planting technique is better than another.

WHEN TO USE?

Ideal for recomposing permanent preservation areas (APPs), helping to bring rural properties into line with the new Forestry Code at a low cost. Landowners who have already used the technique in their areas guarantee that savings can reach 75% compared to conventional reforestation methods. While restoration using traditional seedling planting costs around R$12,000 to R$15,000 per hectare, planting using the muvuca technique costs around R$5,000 per hectare.

The use of muvuca is suitable for large producers, who can use their own agricultural machinery to carry out the planting, but it can also be adapted for small landowners (Conservaçao Internacional - Brazil).

GENERATING JOBS AND INCOME

It is also worth mentioning that in addition to reforestation and the benefits to rural producers, new jobs have also been created. One example of this is seed collectors, who generate income. When the process of reforestation by muvuca is done correctly, the seed collectors are registered in advance and thus have a very important role within the system.

HOW TO USE IT?

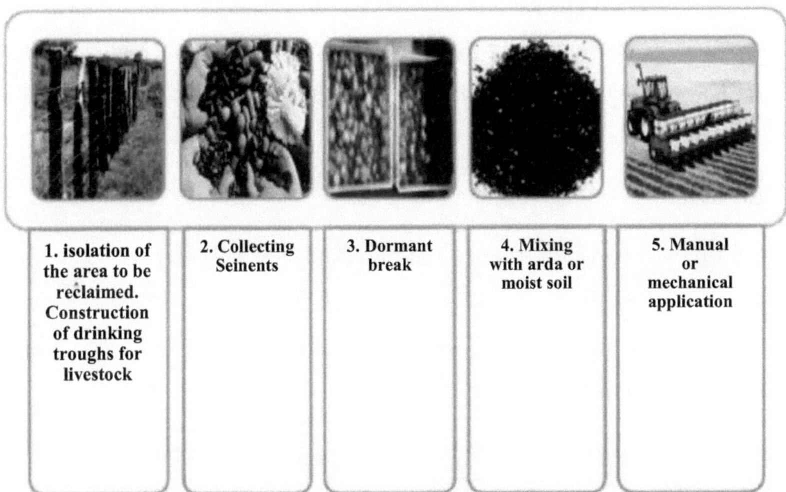

1.   **The** process begins with fencing **off the banks of rivers, lagoons and reservoirs**, so that **cattle** no longer have direct access to the water and no longer cause impacts such as **soil compaction** and **bank erosion.** Drinking troughs built in the middle of the pastures, with controlled access and rotating use, provide for the animals' watering.

2.   **Seeds** should be **collected** from as many species as possible that occur in and around the area. It is **not advisable** to collect seeds from species with aggressive characteristics, especially exotic ones.

3.   Some species find it difficult to germinate using only the natural conditions offered by the environment. To overcome this obstacle, some **seed treatments** should be **carried** out to ensure higher germination rates. In general, immersing the seeds in boiling water is effective in overcoming most of the dormancy mechanisms.

4.   For the best and most uniform application, a **mixture of all the seeds and sand or moist soil** should be prepared. This prevents small seeds from getting lost or blown away by the wind.

5.   **Application** can be carried out mechanically if the landowner has agricultural machinery available, or manually if the landowner is a small farmer. Apply to the whole area and whenever possible to connect existing patches, as this ensures a greater chance of effectiveness over long periods of time. If the area is occupied by grasses, harrowing before sowing is necessary, but if the area is used for agriculture, this step can be dispensed with.

SUGGESTION

Add seeds of short-lived legumes to the muvuca: they will serve as green manure. These legumes reduce the reoccupation of the area by the grass through shading, they decompress and incorporate organic matter and nitrogen into the soil, thus reducing the need for intervention in the area.

Examples: Feijaode-porco (guarantees soil coverage from the first to the tenth month) and Feijao guandu (coverage from the II^0 month to the third year).

2.5 Measures to Activate Ecological Succession

Ecological succession plays a very important role in activating and rehabilitating degraded riparian forest environments, because it is through secondary succession that, when a riparian forest suffers some kind of disturbance such as deforestation, fire or natural causes, the colonization of the area takes place. The main characteristic of secondary succession is that it establishes groups of plants over time, changing the ecology of the site and bringing it closer to a stable and structured phase through a series of successional stages.

For Ferretti (2002), Kageyama *et al.* (2002) and Kageyama and Gandara (2000), secondary succession processes can be considered to be the most applicable in degraded environments since it is through this process that species regenerate naturally and it has a set of measures constituted in the natural process and aimed at maximum cost reduction.

According to Martins (2001), the essential factors for secondary succession to take place are: the proximity of forest surpluses and seed banks, the regrowth of shrub and tree species and the intensity of the alteration, since each environment has specific development characteristics and therefore the regeneration process will also occur differently, however, it is necessary to generally exclude the practice of agricultural activities and extinguish degradation in recovery environments.

According to Engel and Parrota (2003), when ecosystems are extremely degraded, the natural regeneration of the environment will not take place, as it would need a long time to happen, but biotic conditions and human needs prevent this development, so secondary succession becomes unfeasible. Therefore, based on Reis *et al.* (2003) and Mariot (2005), nucleation techniques can be used in areas to be restored, as they are efficient in terms of restoration and make it possible to connect areas separated by forest fragmentation using simple, low-cost techniques. According to Reis *et al.* (2003), nucleation is an origin of succession to naturally colonize areas in formation. Reis and Tres (2008) state that the principles of nucleation are options for recovering riparian forests.

Yarranton and Morrison (1974) define nucleation as "an increase in the rate of colonization from a promoting species". This statement can be understood as an increase in the likelihood of different species occupying an environment, through the improvement in environmental quality provided by a particular species, such as the formation of soil by a first tree species which led to the formation of other secondary and colonizing species.

Thus, Bechara (2006) states that nucleation is a different way of restoring environments, as it allows sporadic phenomena to restore natural areas through secondary succession, preserving the natural ecosystem, but with human intervention, providing food and making the area favorable for the development of flora and fauna all year round.

According to Reis *et al* (2003), nucleation techniques refer to the application of certain methods, such as: transposition of soil and leaf litter, transposition of branches, which serve as shelter for fauna, transposition of seed rain, artificial perches and planting seedlings in dense groups. These techniques will be responsible for encouraging regeneration through interaction between organisms. And, according to Reis and Tres (2008), they are methods that involve the soil and organisms from all trophic levels present in the food chain of the ecosystem involved, to introduce conservation and new elements to the original landscape.

Even though nucleation techniques are an option for restoring riparian forests, it is difficult to improve these environments. This is due to rising river levels that remove or bury seeds and litter, preventing the germination of slow-growing species, but encouraging fast-growing ones. It is necessary to take into account the issue of flooding in these areas, as it is a natural aspect and, according to Rodrigues (2000), favors the initial succession groups.

According to Bechara (2006), the application of nucleating techniques creates microhabitats distributed in clusters throughout the restored area which provide shelter, food and reproduction for various species through the process of successional acceleration.

The main consequences of restoring areas through nucleation techniques are: greater soil cover, environmental protection on slopes, improvements in soil structure, infiltration and percolation, recharge of the water table and management of water resources in basins and minimization of soil erosion. All this is possible through relatively low costs and the implementation of nucleating techniques and an increase in the likelihood of the environment being occupied by other species.

Bechara (2006) says that in Brazil, environmental restoration through the application of nucleation techniques is still recent, but studies show that the use of these techniques reduces costs by around 34% when compared to traditional models. In the state of Minas Gerais, the implementation of such techniques would generate benefits, since there are many areas of riparian forest to be recovered, especially along watercourses. The areas in need of recovery in Minas Gerais are found on small and medium-sized private properties, so these owners can acquire information through this manual to carry out recovery activities in degraded areas.

The study of degraded areas should take place over the long term, so that techniques can be established to improve the reversal of deforestation and promote the integration of cultivable areas, restoring the ecological and functional connectivity of the area (SIGAM/SMA, 2011). However, several authors state that the techniques should not exceed 5% of the restored area, so that functionality is exercised and what is left receives influence from nuclei through secondary succession and the natural conditions of the environment, also taking into account biological and physical issues and the degree of deficiency of the environment. Some of the main nucleation techniques for restoring degraded environments are explained below.

2.5.1.   Piles of twigs or stones

According to Bechara (2006) and Reis and Très (2007), this technique refers to the formation of shaded artificial shelters with low temperatures, high humidity and a high organic matter content. It should act as a perch and thus improve the soil and generate favorable conditions for planting, growth and development of seeds that are adapted to this type of environment, generated from the accumulation of branches. The beds used will be formed from wood, sawmill waste or any leftover woody material from the environment.

Reis *et al.* (2003) states that this nucleation technique creates biodiversity nuclei that are important for the secondary successional process and, according to Espindola *et al.* (2006), the microclimate that is formed in these nuclei makes them act as artificial refuges for the fauna. Reis and Tres (2009) point out that the diversified micro-habitat that is formed also extends to nearby areas as it is expected that the seeds will arrive more easily in the environment to be recovered.

In areas degraded by the removal of topsoil, the main concern is to replenish organic matter so that microorganisms can make nutrients available to colonizing plants, but this organic matter is not always available in areas close to the degraded ones. That's why this nucleation technique becomes efficient, as it manages to get organic matter to form in the areas. The seeds will be carried to the branches by the dispersing species, germinate and occupy the area, in a way rescuing the flora and fauna. The wind will also bring seeds and, as the soil will be suitable for germination, the population will increase.

2.5.2.   Laying perches

In this nucleation technique, the main method to be implemented is the use of perches. These perches, according to Bechara (2006) and Reis and Tres (2007), can be either dry or live. Dry perches act as a landing surface for the birds and provide a resting and foraging structure for them. They should have large canopies and as many dry branches as possible. Live perches, on the other hand, act like living trees, attracting animals that behave in different ways. According to Tres *et al.* (2011), these two types of perch are useful for capturing seeds, as they can attract birds and bats, which are the most efficient animals for transporting and dispersing seeds, especially to degraded areas, and as they land casually on the perches, they leave seeds that adapt and grow. These perches can be set up in different ways and with different functions so that they attract mainly bats and birds. Reis *et al.* (2003) say that these animals accelerate the successional process and the nuclei they generate act as a source of diversity in the initial secondary succession.

Reis and Tres (2009) argue that perches are the best option as a restoration technique, as they are gaining ground and showing their nucleation capacity as ecological stepping stones, generating connectivity between the degraded area and nearby fragments through ecological flows via the implantation of dry or live perches. The biological structures undergo visible changes, since the attraction of consumers to the area acts to facilitate the formation of a new trophic chain, which will become food for secondary consumers and thus generate a new landscape and diversity in the environment.

2.5.3. Soil transpositions

Soil transposition is a nucleation technique that consists of restoring the soil and helping to improve the micro, meso and macro fauna and flora present in the soil, such as seeds, propagules, fungi, bacteria, algae, etc. According to Reis and Vieira (2009), it takes place through the formation of small nuclei containing a seed bank in degraded areas. Reis and Tres (2007) state that transpositions should take place in surface portions of 1 m^2 of soil from conserved areas to nearby degraded areas, with the aim of introducing species that show grouped behavior in nature. Reis *et al.* (2003) states that the soil transposition technique has a high probability of restoring new colonization to an area.

According to Bechara (2006), this nucleation technique promotes the irradiation of propagules through the organism's dispersal capacity. Colonization generates a food supply for consumers and, in order to remain in the community, the species must adapt to climatic and seasonal variations, which is why the pioneers in the successional stages should be the microorganisms that are least sensitive to temperature changes, and only then should the more sensitive macroorganisms be implanted.

Flows of organisms help restore life forms in degraded areas and nearby landscapes. Re-establishing connectivity between them is fundamental for development. According to Reis *et al.* (2003) it is recommended that soils with different successional levels be used in order to have a variation of organisms in the restored ecosystem.

Bechara (2006) points out that soil transposition has a high effect and great potential for re-establishing connectivity between organisms in an ecosystem. This enables the insertion of new forms of life and is important in restoration activities.

2.5.4. Direct sowing

According to Reis and Tres (2008), the direct seeding technique consists of sowing seeds manually or mechanized into the soil in order to create initial cover. This type of technique is indicated when the land is exposed and without vegetation cover after a cleaning activity, for example. It is appropriate to use grass, leguminous and cruciferous species, as these will quickly cover, restructure and develop the soil through a root system that will control the erosion process. The leguminous plants will be responsible for fixing nitrogen, while the crucifers will produce green mass and prevent the soil from breaking up due to their pivoting roots. In order to achieve the best results, there should be a combination of the species mentioned, depending on the time of year and the availability of seeds on the market.

Sowing in declining areas should use the process of plant recomposition, which consists of planting annual and exotic herbaceous species associated with the planting of native seedlings with harbor essence. After the seeds are sown in this type of soil, they should be lightly graded. However, when the slope is steep, such as on embankments or road cuts, the process of hydro-seeding will be used, which has been used and generated very positive results.

Hydroseeding consists of applying an aqueous-pasty mixture containing organic and mineral fertilizers, cellulose or moisture-conserving shredded paper (*acetamulch*), an adhesive to fix the cellulose and grass and legume seeds, placed last to reduce breakage due to mechanical friction. O planting should be carried out in rainy seasons so that there is no watering. However, before hydroseeding is implemented, the slope must be spaded, which consists of making furrows or small pits. These can be horizontal along the slope, with a distance of 20 to 30 cm between each furrow and each pit 3 to 5 cm wide; or they can be distributed irregularly on the surface of the slope, 10 cm apart; in both cases there should be a depth of 5 cm in all the pits to improve seed penetration and moisture regulation.

The species indicated for use in seeding or hydro-seeding techniques are:

- *Cibopogom citratus* (lemon grass);

- *Paspalum notatum* (batatal grass);

- *Cajanus cajan* (guandu bean);

- *Desmodiumspp* (desmodium);

- *Stylozanthesspp* ;

2.5.5 Planting Seedlings in Anderson Groups

This type of technique, explained by Anderson in 1953, consists of planting seedlings produced in forest nurseries in order to create nuclei capable of attracting a greater biological diversity to degraded areas and making the nucleation process efficient, while the quality of the genetic material to be planted is also important. The seedlings should be arranged homogeneously or heterogeneously and 3.5 or 13 seedlings should be planted.

Reis *et al.* (2006) explains that these nuclei should be made up of five seedlings arranged in a "+" shape, spaced 0.5 x 0.5m apart, with four seedlings at the edges and one in the center. According to Espindola *et al.* (2006), the central seedlings benefit in terms of height development and the lateral ones in terms of branching. However, this technique can be inappropriate, as it fixes the successional process for too long and promotes the development of only the planted species. This procedure skips the stages of succession and gives importance only to the structure of the forest, forgetting about the formation of biodiversity in the environment.

This is why Kageyama and Gandara (2000) propose that these cores should be made up of different species such as herbs, shrubs, lianas and trees. The species used are early maturing, can flower and bear fruit very quickly and attract a variety of organisms to the cores, generating the capacity for adaptation and reproduction of other living beings. For this reason, it is recommended that the nucellos provide a diversified food supply throughout the year for all the species in the environment.

According to Damasceno (2005), monitoring carried out in these restored areas showed that this technique does not guarantee the self-sustainability of the environment, i.e. new species did not enter the restored environment and the environment showed low diversity and floristic complexity;

Espindola *et al.* (2006) recommend that the seedlings receive care such as fertilizing and hilling until they form a shaded core that allows the growth of spciophytic species.

Printed by Books on Demand GmbH, Norderstedt / Germany